MARS BOUND

The Next Frontier of Human Exploration

Geoffrey Zachary

CONTENTS

Survival and Exploration

Mars Bound: The Next Frontier of Human Exploration

PART I: PREPARING FOR MARS

CHAPTER 1: THE RED PLANET: A BRIEF INTRODUCTION TO MARS

Introduction:
Mars, often referred to as the Red Planet, has captured the imagination of humans for centuries. With its rusty hue and mysterious allure, it has become the next frontier of human exploration. In this chapter, we will delve into the fascinating world of Mars, exploring its physical characteristics, historical significance, and the allure it holds for scientists and adventurers alike.

Understanding Mars:
Mars is the fourth planet from the Sun in our solar system and is often described as Earth's neighbouring planet. It has a diameter of approximately 6,779 kilometres (4,212 miles), about half the size of Earth, and a surface area equivalent to the combined landmasses of Earth's continents. Mars is known for its distinctive reddish appearance, caused by iron oxide, or rust, covering its surface.

Geological Features:
Mars boasts a diverse range of geological features that have intrigued scientists for decades. Its surface is marked by vast plains, towering volcanoes, deep canyons, and polar ice caps.

One of its most famous features is Olympus Mons, the largest volcano in the solar system, standing at a staggering height of 25 kilometres (16 miles). Valles Marineris, a system of canyons stretching over 4,000 kilometres (2,500 miles) long, is another captivating geological wonder.

Water on Mars:
One of the key elements driving human interest in Mars is the possibility of water, a fundamental ingredient for life as we know it. Over the years, multiple missions have revealed evidence of water in various forms on Mars, including ice at the poles and subsurface water reservoirs. The discovery of ancient riverbeds and sedimentary layers suggests that Mars was once a wetter and potentially habitable planet.

Historical Significance:
Mars has long captivated the human imagination. Ancient civilizations, such as the Egyptians and Greeks, observed its movements in the night sky and associated it with gods and myths. In more recent history, scientists and astronomers have conducted observations and sent robotic missions to Mars, uncovering its secrets, and paving the way for future exploration.

Mars Missions:
The quest to explore Mars has led to a series of successful missions, each contributing to our understanding of the planet. In 1971, the Soviet Union's Mars 3 became the first spacecraft to land on Mars, albeit for a brief period. The Viking missions, launched by NASA in the 1970s, provided valuable data about the Martian atmosphere and surface. In recent years, the Curiosity rover has been roving the Martian landscape, collecting data about its geology and potential habitability.

The Allure of Mars:
Mars has a unique allure that continues to inspire scientists, explorers, and dreamers around the world. Its potential as a future destination for human settlement, its similarities and

differences to Earth, and the unanswered questions about its past and potential for life make it an enticing subject of study and exploration.

Real-Life Inspiration:
The exploration of Mars has not only been limited to scientific missions. It has also captured the public's imagination, inspiring countless books, movies, and dreams of space travel. The successful landing of the Perseverance rover in February 2021, with its breath-taking images and the historic Ingenuity helicopter flights, has reignited excitement and curiosity about Mars.

Conclusion:
As we conclude this introductory chapter, it becomes evident that Mars is not just a distant planet but a world of intrigue and fascination. Its unique geological features, historical significance, and potential for life make it a compelling destination for future human exploration. In the chapters that follow, we will delve deeper into the challenges, advancements, and aspirations associated with the next frontier: Mars.

CHAPTER 2: THE HISTORY OF MARS EXPLORATION: FROM ANCIENT OBSERVATIONS TO ROBOTIC MISSIONS

Introduction:
The exploration of Mars has a rich history that spans centuries, from ancient observations to modern-day robotic missions. In this chapter, we will delve into the fascinating journey of human exploration on the Red Planet, tracing the footsteps of early astronomers, pioneers, and the remarkable technological advancements that have brought us closer to understanding the mysteries of Mars.

Ancient Observations:
Mars has fascinated humans for millennia. Ancient civilizations such as the Egyptians, Greeks, and Chinese observed the planet's distinct reddish appearance and its movement across the night sky. They associated Mars with deities and incorporated it into their mythology. These early observations laid the foundation for our understanding of the Red Planet.

Early Attempts and Telescopic Observations:
In the 17th and 18th centuries, with the invention of telescopes, astronomers like Giovanni Schiaparelli and Percival Lowell made significant contributions to the study of Mars. Schiaparelli's observations of "canali" on Mars sparked debate and fuelled imagination, although the term was later found to be a misinterpretation of the Italian word for "channels."

First Robotic Missions:
The space age brought about a new era of Mars exploration. The Soviet Union's Mars program in the 1960s and early 1970s aimed to be the first to send robotic spacecraft to the Red Planet. Although some missions failed, the Mars 3 mission successfully landed on Mars in 1971, making it the first successful soft landing on another planet.

Viking Missions and the Search for Life:
In the 1970s, NASA's Viking missions became a pivotal moment in Mars exploration. Viking 1 and Viking 2 were the first spacecraft to land on Mars and conduct experiments in search of life. Although the results were inconclusive, these missions provided valuable data on the planet's soil composition and atmospheric conditions.

Rovers: Spirit, Opportunity, and Curiosity:
The turn of the millennium brought new opportunities for Mars exploration. NASA's Mars Exploration Rovers, Spirit, and Opportunity were sent to the planet in 2003. These rovers exceeded their expected mission durations and provided valuable insights into Martian geology and the potential for past habitability. The subsequent Curiosity rover, launched in 2011, continues to explore the planet's surface, analysing samples and searching for signs of ancient microbial life.

Recent Advancements:
In recent years, Mars exploration has reached new heights. The successful landing of the Perseverance rover in February 2021

has opened new avenues for scientific exploration. Equipped with state-of-the-art instruments and the Ingenuity helicopter, Perseverance aims to study Mars' geology, search for signs of ancient life, and pave the way for future human missions.

Real-Life Stories of Exploration:
The history of Mars exploration is filled with captivating stories of perseverance and discovery. The story of the Opportunity Rover's survival through a massive dust storm and its remarkable longevity in exploring the Martian surface stands as a testament to human ingenuity and determination. These real-life examples inspire us to push the boundaries of knowledge and explore the unknown.

The Legacy and Future:
The legacy of Mars exploration has laid the foundation for future missions and the eventual goal of human colonization. The wealth of knowledge gained from robotic missions informs our understanding of the planet's geology, atmosphere, and potential for life. As technology advances and international collaborations grow, the future of Mars exploration holds promises of groundbreaking discoveries and significant advancements in our quest to unravel the mysteries of the Red Planet.

Conclusion:
In conclusion, the history of Mars exploration is a testament to human curiosity, ingenuity, and our relentless pursuit of knowledge. From ancient observations to the modern marvels of robotic missions, each step in our journey has brought us closer to understanding Mars and our place in the universe. The stories of early astronomers, pioneering missions, and ongoing discoveries inspire us to continue exploring the next frontier of human exploration: Mars.

CHAPTER 3: MISSION TO MARS: PLANNING AND DESIGNING HUMAN EXPEDITIONS

Introduction:
The dream of sending humans to Mars has captivated the imagination of scientists, engineers, and space enthusiasts around the world. In this chapter, we will explore the intricacies of planning and designing human expeditions to the Red Planet, examining the various challenges, considerations, and potential solutions that pave the way for this monumental endeavour.

The Importance of Mars Missions:
Mars missions have long been a focus of scientific and technological advancement due to the planet's potential for harbouring life and its similarities to Earth. Human expeditions to Mars hold the promise of answering fundamental questions about the existence of extra-terrestrial life, understanding the planet's geological history, and expanding human civilization beyond our home planet.

Mission Objectives and Goals:
A successful mission to Mars requires clear objectives and goals. These may include scientific research, technological demonstrations, resource utilization, and the establishment of a sustainable human presence on the planet. Each objective must be

carefully defined to maximize the scientific and societal benefits of the mission.

The Journey to Mars:
The journey to Mars presents numerous challenges, including the long duration of the trip, the physical and psychological effects of deep space travel on astronauts, and the need for self-sufficiency during the journey. Mission planners must carefully consider propulsion systems, life support systems, radiation protection, and crew well-being to ensure a safe and successful journey.

Landing and Surface Operations:
Landing on Mars poses unique challenges due to its thin atmosphere and variable terrain. The entry, descent, and landing (EDL) phase of the mission requires precise engineering and navigation to ensure a controlled and safe landing. Once on the surface, astronauts will need to establish habitats, conduct scientific experiments, and utilize local resources to sustain their mission.

Life Support Systems:
Maintaining life support systems in the harsh Martian environment is crucial for the survival of astronauts. Systems for generating oxygen, recycling water, and producing food must be designed to operate efficiently and sustainably. Learning from the experiences of long-duration space missions, such as the International Space Station, will inform the development of robust life support systems for Mars missions.

Communication and Mission Control:
Effective communication between Earth and Mars is vital for mission success. Delayed communication, known as the "Mars-Earth latency," requires mission control teams to anticipate and address contingencies without real-time interaction. Advanced communication technologies and protocols will play a crucial role in ensuring efficient information exchange and decision-making.

Risk Management and Mitigation:

Mars missions come with inherent risks, including radiation exposure, psychological stress, and potential equipment failures. Robust risk management strategies, contingency plans, and redundant systems must be implemented to mitigate these risks and ensure the safety of the crew. Learning from previous space missions and conducting thorough simulations and testing will be key to minimizing risks.

Real-Life Examples:
Real-life examples such as NASA's Mars mission architecture, including the proposed Artemis program, and private initiatives like SpaceX's Starship demonstrate the ongoing efforts to plan and design human expeditions to Mars. These initiatives involve collaborations with international partners, technological advancements, and innovative mission concepts that inspire future explorers and push the boundaries of human spaceflight.

Conclusion:
The planning and design of human expeditions to Mars is a complex and multidisciplinary endeavour. It requires meticulous attention to detail, cutting-edge technologies, and a deep understanding of the challenges involved. Through careful planning, collaboration, and continuous innovation, we are edging closer to realizing the dream of sending humans to Mars. By venturing to the Red Planet, we not only expand our scientific knowledge but also push the boundaries of human exploration and inspire future generations to reach for the stars.

CHAPTER 4: ASTRONAUT SELECTION AND TRAINING FOR MARS MISSIONS

Introduction:
As we embark on the journey to send humans to Mars, one of the critical aspects is the selection and training of the astronauts who will undertake this extraordinary mission. In this chapter, we will delve into the rigorous process of astronaut selection, the comprehensive training they undergo, and the qualities and skills required to become a Mars explorer.

The Selection Process:
Selecting astronauts for a Mars mission is a meticulous process that involves a combination of physical, psychological, and intellectual assessments. Space agencies and organizations worldwide establish rigorous criteria to identify individuals who possess the right attributes, including physical fitness, mental resilience, adaptability, teamwork, and problem-solving abilities. The selection process often includes extensive medical evaluations, psychological assessments, interviews, and simulations.

Physical Fitness and Health:

Astronauts bound for Mars must maintain exceptional physical fitness and health to withstand the demanding conditions of space travel and prolonged periods in microgravity. They undergo rigorous physical training regimes, including cardiovascular exercises, strength training, and endurance conditioning. Additionally, they receive specialized medical training to address potential health issues during long-duration space missions.

Psychological Resilience and Teamwork:
The psychological well-being and resilience of astronauts are crucial factors for successful missions to Mars. Isolation, confinement, and extended periods away from loved ones pose significant challenges. Astronauts undergo psychological evaluations and receive training to develop coping mechanisms, stress management techniques, and effective communication skills. Teamwork and collaboration are emphasized to ensure a harmonious and efficient crew dynamic.

Technical and Scientific Expertise:
Astronauts must possess a diverse range of technical and scientific expertise to conduct research, maintain spacecraft, and handle emergencies. They undergo intensive training in systems engineering, space science, robotics, and extravehicular activities (EVAs) to perform tasks such as repairs, maintenance, and scientific experiments on the Martian surface. Their training also includes simulations of various mission scenarios and hands-on experience with mission-specific equipment.

Real-Life Examples:
Real-life examples of astronaut selection and training programs, such as those conducted by NASA, ESA, and other space agencies, offer insights into the comprehensive nature of preparing individuals for Mars missions. For instance, NASA's Astronaut Candidate Program includes extensive physical and psychological evaluations, rigorous academic coursework, survival training, and simulated space missions. These programs ensure that astronauts are well-prepared and equipped with the necessary

skills to thrive in the challenging environment of space.

Inspiring Stories:
The chapter may highlight inspiring stories of astronauts who have gone through the selection and training process and have embarked on space missions. For example, the selection of the "Mercury Seven" astronauts for NASA's Mercury program in the 1960s marked the beginning of human space exploration. Their bravery, resilience, and dedication to pushing the boundaries of human exploration continue to inspire future generations of astronauts.

Conclusion:
Selecting and training astronauts for Mars missions is a meticulous and multidimensional process that combines physical fitness, psychological resilience, technical expertise, and scientific acumen. The individuals chosen for these missions represent the epitome of human potential and embody the spirit of exploration and discovery. Through rigorous selection and comprehensive training, we prepare these remarkable individuals to venture into the unknown, inspiring generations to reach for the stars and realize the dream of human exploration on Mars.

PART II: THE JOURNEY TO MARS

CHAPTER 5: LAUNCHING TO MARS: ROCKETS, PROPULSION SYSTEMS, AND TRAJECTORY CALCULATIONS

Introduction:
The journey to Mars requires overcoming significant challenges, and a crucial aspect of this endeavour is the process of launching spacecraft from Earth and propelling them towards the Red Planet. In this chapter, we will explore the intricate world of rockets, propulsion systems, and trajectory calculations involved in launching missions to Mars.

Rockets and Propulsion Systems:
Rockets serve as the primary means of transportation for space missions, including those destined for Mars. They utilize powerful propulsion systems to generate enough thrust to escape Earth's gravitational pull and enter orbit. Two main types of rockets are commonly used: expendable rockets, which are single-use

and discarded after launch, and reusable rockets, which can be recovered and used for multiple missions, reducing costs significantly. Examples of notable rockets include the Falcon 9 by SpaceX and the Delta IV by United Launch Alliance.

Rocket Staging and Engine Operation:
Rockets consist of multiple stages, each equipped with its engine. The initial stages provide the initial thrust to lift the spacecraft off the ground, while the upper stages continue the journey towards Mars. Propellant fuels, such as liquid oxygen and liquid hydrogen, are commonly used in rocket engines due to their high energy efficiency. The engines operate on the principles of combustion and Newton's third law of motion, expelling exhaust gases at high speeds to generate thrust.

Trajectory Calculations:
Precise trajectory calculations are crucial for ensuring that spacecraft reach Mars accurately and efficiently. Mission planners and engineers utilize complex mathematical models and simulations to determine the optimal trajectory, considering variables such as launch window, orbital dynamics, gravity assists, and fuel efficiency. These calculations are essential to minimize travel time and ensure the spacecraft's safe arrival at Mars.

Real-Life Examples:
Real-life examples showcase the advancements and challenges in launching missions to Mars. For instance, NASA's Mars missions, including the Mars rovers Curiosity and Perseverance, were launched using powerful rockets such as the Atlas V and the Mars Science Laboratory. The successful launches and subsequent missions demonstrate the effectiveness of the propulsion systems and trajectory calculations employed in these missions.

Inspiring Stories:
Inspiring stories of successful rocket launches, and Mars missions can captivate readers and highlight the incredible achievements

of human exploration. One example is the historic landing of NASA's Mars Science Laboratory, also known as the Curiosity rover, in 2012. This mission showcased the culmination of years of planning, engineering, and precise calculations, leading to the rover's safe arrival on Mars. Such stories serve as a testament to human ingenuity and the pursuit of knowledge.

Conclusion:
Launching missions to Mars requires a deep understanding of rockets, propulsion systems, and trajectory calculations. The development of advanced rockets, the efficient operation of engines, and the precise calculations of trajectories are essential for successful Mars missions. Real-life examples and inspiring stories demonstrate the significant progress made in this field, inspiring future generations of engineers and space explorers to push the boundaries of human exploration. By mastering the art of launching spacecraft to Mars, we inch closer to realizing our dream of becoming an interplanetary species.

CHAPTER 6: LIVING IN TRANSIT: CHALLENGES AND SOLUTIONS FOR LONG-DURATION SPACE TRAVEL

Introduction:
As we embark on the journey to Mars, one of the greatest challenges we face is the long-duration space travel required to reach our destination. In this chapter, we will explore the physical and psychological challenges astronauts will encounter during extended space missions, as well as the innovative solutions being developed to ensure their well-being and productivity throughout the journey.

Physical Challenges:
Extended periods of weightlessness can have profound effects on the human body. Bone density loss, muscle atrophy, cardiovascular changes, and weakened immune systems are some of the physical challenges that astronauts experience during space travel. To mitigate these effects, exercise regimes, including resistance training and cardiovascular workouts, are implemented to maintain muscle and bone health. Additionally,

advanced medical monitoring and diagnostic systems are utilized to closely monitor astronauts' health and intervene when necessary.

Psychological Challenges:
Isolation, confinement, and the psychological stressors of long-duration space travel pose significant challenges for astronauts. Being confined in a small spacecraft for months or even years can lead to feelings of boredom, loneliness, and homesickness. To address these challenges, astronauts undergo rigorous psychological training and are provided with support systems, including regular communication with mission control and their families. Additionally, recreational activities, such as virtual reality simulations and personal hobbies, are encouraged to maintain mental well-being.

Sleep and Circadian Rhythm:
Maintaining a healthy sleep schedule and circadian rhythm is crucial for the overall well-being and performance of astronauts. In space, where the concept of day and night is altered, astronauts may experience sleep disturbances and disruptions to their internal body clocks. To address this, spacecraft are equipped with carefully designed lighting systems that mimic natural sunlight and darkness to regulate astronauts' sleep-wake cycles. Additionally, personalized sleep environments and relaxation techniques are employed to promote quality sleep.

Nutrition and Food Systems:
Providing adequate nutrition for astronauts during long-duration space travel is essential for their health and performance. Food must be carefully selected and preserved to meet nutritional requirements while remaining safe and palatable. Advanced food systems, such as packaged meals, nutrient-dense foods, and hydroponic gardens, are employed to ensure a balanced diet. Moreover, advancements in food processing techniques, such as freeze-drying and vacuum-sealing, enable the preservation of food for extended periods.

Real-Life Examples:
Real-life examples, such as the International Space Station (ISS), serve as testbeds for studying and developing solutions for long-duration space travel. Astronauts aboard the ISS experience many of the challenges discussed in this chapter and contribute valuable insights into mitigating their effects. The knowledge gained from these missions informs the planning and preparation for future Mars missions.

Inspiring Stories:
Inspiring stories of astronauts who have spent extended periods in space, such as Scott Kelly's year-long mission on the ISS, highlight the resilience and adaptability of the human spirit. These stories demonstrate the determination and dedication required to overcome the physical and psychological challenges of long-duration space travel. They inspire future astronauts and space explorers to push the boundaries of human endurance and explore the vastness of our universe.

Conclusion:
Long-duration space travel presents significant physical and psychological challenges for astronauts. Through innovative solutions and a deep understanding of human physiology and psychology, scientists and engineers are developing strategies to mitigate these challenges. The lessons learned from real-life examples and inspiring stories of space exploration inspire us to continue pushing the boundaries of human endurance and pave the way for future generations to embark on the extraordinary journey to Mars.

CHAPTER 7:
MAINTAINING HEALTH AND WELL-BEING IN SPACE: MEDICAL CONSIDERATIONS FOR MARS MISSIONS

Introduction:
As humans venture further into space and prepare for the goal of reaching Mars, ensuring the health and well-being of astronauts becomes paramount. The harsh space environment presents unique challenges that must be addressed to mitigate potential health risks during long-duration Mars missions. In this chapter, we will explore the medical considerations and innovative strategies employed to safeguard the physical and mental health of astronauts on their journey to the Red Planet.

Physical Health Considerations:
Extended exposure to microgravity, radiation, and confined spaces can have profound effects on the human body. Astronauts may experience cardiovascular deconditioning, muscle and bone loss, and changes in the immune system. To counteract these effects, rigorous exercise regimes, including resistance training and aerobic exercise, are implemented to maintain muscle

and bone strength, cardiovascular fitness, and overall health. Additionally, nutritional strategies are employed to ensure astronauts receive the necessary nutrients to support their physiological needs.

Radiation Protection:
Space is filled with radiation from various sources, including solar particles and cosmic rays, which pose a significant risk to human health. Shielding technologies and spacecraft design considerations help minimize astronauts' exposure to radiation. Additionally, monitoring systems and dosimeters are utilized to measure radiation levels and inform necessary precautions. Understanding the long-term effects of radiation exposure and developing effective mitigation strategies are critical for the success of Mars missions.

Telemedicine and Remote Healthcare:
In the isolated and extreme environment of space, access to immediate medical care and expertise is crucial. Telemedicine, which involves remote medical consultations and diagnostics, allows astronauts to receive real-time medical guidance from experts on Earth. Advanced medical equipment and diagnostic technologies enable astronauts to perform medical procedures and tests onboard the spacecraft under the guidance of medical professionals. This telemedicine capability ensures timely and appropriate medical interventions during emergencies or routine health assessments.

Psychological Well-being:
The psychological well-being of astronauts is equally important as their physical health during long-duration space missions. The isolation, confinement, and distance from loved ones can lead to feelings of loneliness, stress, and anxiety. Psychological support systems, such as regular communication with family and friends, counselling services, and recreational activities, are implemented to help astronauts cope with the challenges of space travel. Psychological training and resilience-building exercises equip

astronauts with the tools to maintain mental well-being and perform effectively in a high-stress environment.

Real-Life Examples:
Real-life examples, such as the International Space Station (ISS) and long-duration space missions, provide valuable insights into medical considerations for Mars missions. Astronauts who have spent extended periods in space, like Scott Kelly and Peggy Whitson, have contributed to our understanding of the physical and psychological effects of space travel. Their experiences inform the development of medical protocols, exercise regimes, and psychological support systems for future Mars missions.

Inspiring Stories:
The stories of astronauts overcoming physical and mental challenges during their space missions serve as inspiration for future astronauts and the public. These stories highlight the resilience, determination, and adaptability of the human spirit. They demonstrate the power of human ingenuity and innovation in overcoming the inherent risks of space travel and inspire us to push the boundaries of exploration and scientific progress.

Conclusion:
Maintaining the health and well-being of astronauts during Mars missions is a complex undertaking. Through a comprehensive understanding of the physiological and psychological effects of space travel, innovative medical technologies, and support systems, we can ensure the safety and success of future missions. Real-life examples and inspiring stories of astronauts embarking on incredible journeys further ignite our curiosity and drive to explore the mysteries of space while safeguarding the health and well-being of those who dare to venture to the Red Planet.

CHAPTER 8: SUSTAINABILITY IN SPACE: FOOD, WATER, AND LIFE SUPPORT SYSTEMS

Introduction:
As humans embark on the journey to Mars, the sustainability of life support systems becomes crucial. The ability to produce and maintain a sustainable supply of food, water, and breathable air is essential for the survival and well-being of astronauts during long-duration missions. In this chapter, we will explore the challenges, innovative solutions, and real-life examples related to sustainability in space.

Challenges of Sustainability in Space:
Space missions require self-sufficiency in terms of essential resources. The limited availability of water, lack of atmospheric oxygen, and the absence of fertile soil pose significant challenges for sustaining life beyond Earth. Moreover, the need for lightweight and compact systems further adds to the complexity of developing sustainable life support systems for space travel.

Food Production in Space:
To address the challenge of food sustainability, space agencies and researchers have been exploring different methods of

growing food in space. Hydroponics, aeroponics, and other soil-less cultivation techniques have shown promise in enabling astronauts to grow fresh produce aboard spacecraft. NASA's Veggie experiment, for example, successfully grew lettuce on the International Space Station (ISS), showcasing the potential for sustainable food production in microgravity environments.

Water Management and Recycling:
Water is a critical resource in space, and its conservation and recycling are paramount for long-duration missions. Systems for water purification and recycling are employed to minimize water consumption and ensure the availability of safe drinking water. The Water Recovery System on the ISS is an example of advanced water recycling technology that converts wastewater into potable water for astronauts' consumption.

Air Revitalization and Oxygen Generation:
Maintaining breathable air is essential for the survival of astronauts. Life support systems employ air revitalization technologies that remove carbon dioxide and replenish oxygen levels. The Environmental Control and Life Support System (ECLSS) on the ISS utilizes processes such as electrolysis to generate oxygen from water and carbon dioxide removal systems to maintain a suitable atmosphere for astronauts.

Real-Life Examples:
Real-life examples, such as the ISS, provide valuable insights into the sustainability of life support systems in space. The continuous operation of the ISS for over two decades showcases the successful management of resources, including food, water, and air, sustainably. Lessons learned from these missions inform the development of future sustainable life support systems for Mars missions.

Innovative Solutions:
Innovative technologies are being developed to enhance sustainability in space. 3D printing, for instance, offers the

potential to produce spare parts and tools on demand, reducing the need for extensive supply missions. Closed-loop systems that recycle waste and extract valuable resources, such as water and nutrients, from organic waste, are also being explored. These advancements contribute to the long-term sustainability of human presence in space.

Inspiring Stories:
The journey towards sustainability in space is not without its challenges, but the progress made so far inspires hope for a future where humans can thrive beyond Earth. The dedication and ingenuity of scientists, engineers, and astronauts who work tirelessly to develop sustainable life support systems serve as a testament to the indomitable human spirit and our ability to adapt and overcome obstacles.

Conclusion:
Achieving sustainability in space is crucial for the success of long-duration missions to Mars and beyond. Through innovative technologies and resource management strategies, we are making significant progress in ensuring the availability of food, water, and breathable air for astronauts. Real-life examples from the ISS and ongoing research efforts inspire us to continue pushing the boundaries of sustainability in space exploration. By developing and implementing sustainable life support systems, we pave the way for a future where humanity can sustainably thrive in the harsh environments of space.

PART III: LANDING AND ESTABLISHING A PRESENCE ON MARS

CHAPTER 9: ENTRY, DESCENT, AND LANDING: OVERCOMING THE CHALLENGES OF THE MARTIAN ATMOSPHERE

Introduction:
As humans venture to Mars, one of the most critical and challenging phases of the mission is the entry, descent, and landing (EDL) onto the Martian surface. The thin atmosphere and unique conditions of Mars pose significant challenges that must be overcome to ensure a safe and successful landing. In this chapter, we will explore the intricacies of EDL and the innovative technologies and strategies employed to overcome the challenges of the Martian atmosphere.

The Martian Atmosphere:
The Martian atmosphere is approximately 100 times thinner than Earth's atmosphere, consisting mainly of carbon dioxide. This thin atmosphere provides little aerodynamic drag, making

it difficult to decelerate spacecraft during entry. Additionally, the low atmospheric pressure poses challenges for parachute deployment and the stability of landing systems.

Aerocapture and Atmospheric Entry:
To overcome the challenges of entering the Martian atmosphere, spacecraft often utilize a technique called aerocapture. Aerocapture involves using the Martian atmosphere to slow down the spacecraft and place it into orbit around Mars. This technique minimizes the need for propellant and reduces the overall mission cost.

Heat Shield and Thermal Protection:
During atmospheric entry, spacecraft experience intense heating due to the friction between the vehicle and the thin Martian atmosphere. Heat shields made of advanced materials, such as carbon composites or ceramic tiles, are used to protect the spacecraft and its payload from extreme temperatures. The heat shield absorbs and dissipates the heat, ensuring the integrity of the spacecraft.

Parachute Deployment:
Once the spacecraft has decelerated sufficiently, a parachute is deployed to further slowdown the descent. Parachutes provide the necessary drag to reduce the speed of the spacecraft, allowing for a controlled descent towards the Martian surface. The Viking landers and the recent Perseverance rover mission are examples of successful parachute deployments during Mars missions.

Powered Descent and Landing Systems:
As the spacecraft approaches the Martian surface, additional systems are employed to ensure a safe landing. Retro propulsion, or the use of rocket engines, is often utilized for the final phase of descent. The engines provide the necessary thrust to slow down the spacecraft and achieve a soft landing. The SpaceX Star ship's landing capability, demonstrated during recent test flights, showcases the potential of powered descent for Mars missions.

Real-Life Examples:
Real-life examples of successful Mars landings, such as the NASA Mars rovers Spirit, Opportunity, Curiosity, and Perseverance, highlight the engineering achievements and technological advancements made in EDL systems. These missions have provided valuable data and insights into the challenges and complexities of landing on Mars, paving the way for future missions.

Future Technologies and Concepts:
To further improve landing capabilities on Mars, researchers and engineers are exploring innovative technologies and concepts. These include supersonic retro propulsion, precision landing systems, and inflatable decelerators. These advancements aim to enhance the precision, reliability, and efficiency of Mars landings, enabling more ambitious missions in the future.

Inspiration from EDL:
The successful entry, descent, and landing on Mars serve as sources of inspiration and awe. They demonstrate the ingenuity and determination of the human spirit to explore and conquer new frontiers. The challenges faced during EDL highlight the collaboration and expertise of scientists, engineers, and space agencies worldwide, inspiring future generations to push the boundaries of space exploration.

Conclusion:
Entry, descent, and landing are critical stages in the journey to Mars. Overcoming the challenges posed by the Martian atmosphere requires advanced technologies, meticulous planning, and a deep understanding of the physics involved. Through innovative approaches and the lessons learned from previous missions, we continue to improve our ability to safely land spacecraft on Mars. The successful landings and ongoing advancements in EDL systems ignite our curiosity and fuel our desire to explore the Red Planet further.

CHAPTER 10: BUILDING THE FIRST MARTIAN HABITAT: DESIGNING FOR SURVIVAL AND EXPLORATION

Introduction:
As humans set their sights on Mars, the establishment of a sustainable habitat becomes crucial for long-term missions and the exploration of the Red Planet. In this chapter, we delve into the intricacies of designing and constructing the first Martian habitat, considering the challenges of the Martian environment and the requirements for human survival and exploration.

Understanding the Martian Environment:
The Martian environment poses numerous challenges that must be addressed in habitat design. Mars experiences extreme temperatures, with average surface temperatures around -80 degrees Fahrenheit (-62 degrees Celsius). The planet's thin atmosphere lacks oxygen and protection from harmful solar radiation, requiring habitats to provide adequate shielding and life support systems.

Structural Design and Materials:
To withstand the harsh Martian environment, habitats must be built using durable and resilient materials. Inflatable structures, such as those used in the Bigelow Expandable Activity Module (BEAM) on the International Space Station, offer lightweight and compact solutions for transporting and deploying habitats on Mars. Additionally, using locally available resources, such as Martian soil (regolith), can reduce the reliance on Earth-sourced materials and decrease mission costs.

Life Support Systems:
Creating a self-sustaining habitat on Mars necessitates the integration of advanced life support systems. These systems must provide a breathable atmosphere, manage waste, and water recycling, and offer sufficient food production capabilities. Technologies like closed-loop life support systems, hydroponics, and algae cultivation can contribute to the sustainability of a Martian habitat.

Radiation Shielding:
The thin atmosphere of Mars provides minimal protection against cosmic radiation, making radiation shielding a critical consideration in habitat design. Employing radiation-absorbing materials, such as water-filled walls or regolith shielding, can help mitigate the health risks associated with long-term radiation exposure.

Energy Generation and Storage:
Mars receives less solar energy compared to Earth due to its greater distance from the sun and its thin atmosphere. Therefore, efficient energy generation and storage systems are essential for sustaining a Martian habitat. Solar panels, coupled with energy storage technologies like batteries or fuel cells, can ensure a reliable power supply during periods of low sunlight or dust storms.

Real-Life Examples:

Real-life examples of habitat design and construction, such as NASA's Mars Habitat 1:1 scale mock-up and the Mars Society's Mars Desert Research Station, showcase the innovative approaches taken to simulate and test habitat technologies and concepts. These simulations provide valuable insights into the challenges and opportunities presented by the Martian environment, informing future habitat designs.

Exploration and Expansion:
The first Martian habitat will serve as a base for scientific research, exploration, and the preparation for further colonization efforts. It will house astronauts during their missions and provide them with the necessary resources and support systems. The lessons learned from the initial habitat will inform the design and construction of larger, more permanent settlements as humans expand their presence on Mars.

Inspiring Future Generations:
The development and construction of the first Martian habitat represent a significant milestone in human history and a testament to our ingenuity and determination. It inspires future generations to pursue careers in science, engineering, and space exploration, fostering a passion for innovation and pushing the boundaries of what is possible.

Conclusion:
Building the first Martian habitat is an ambitious undertaking that requires careful planning, innovative design, and cutting-edge technologies. The habitat will be a testament to human resilience and adaptability in the face of extreme environments. By addressing the challenges of the Martian environment and ensuring the sustainability and safety of the habitat, we can pave the way for extended human presence on Mars and unlock the mysteries of the Red Planet.

CHAPTER 11: MARS IN 3D: GEOLOGY, TOPOGRAPHY, AND EXPLORATION SITES

Introduction:
In the quest to explore Mars, understanding its geology, topography, and potential exploration sites is crucial for selecting landing sites and conducting scientific investigations. In this chapter, we delve into the 3D representation of Mars, exploring its diverse geological features, identifying key exploration sites, and discussing their significance in unravelling the planet's past and potential for future human missions.

Mapping the Martian Surface:
Over the years, spacecraft orbiting Mars has provided us with a wealth of data and imagery, allowing us to create detailed maps and 3D models of the Martian surface. These maps provide valuable insights into the planet's geology, revealing ancient riverbeds, impact craters, volcanoes, and even evidence of possible past or present water sources. By examining these features, scientists can gain a better understanding of Mars' history and its potential to support life.

Valles Marineris: The Martian Grand Canyon:
One of the most prominent geological features on Mars is Valles Marineris, a vast canyon system that stretches over 4,000

kilometres (2,500 miles) across the planet's surface. It is nearly ten times longer and five times deeper than the Grand Canyon on Earth. Valles Marineris offers a window into Mars' geological past, with its layered walls revealing the planet's ancient history of tectonic activity and erosion.

Olympus Mons: The Largest Volcano in the Solar System:
Mars is also home to Olympus Mons, the largest volcano in the solar system. Standing at a height of 22 kilometres (13.6 miles), it is nearly three times the height of Mount Everest. This colossal shield volcano showcases the planet's volcanic activity and provides scientists with insights into the planet's geologic processes and potential for past or present habitability.

Exploration Sites: Gale Crater and Jezero Crater:
Two notable exploration sites on Mars are Gale Crater and Jezero Crater. Gale Crater is home to the Curiosity rover, which has been exploring the Martian surface since 2012. The crater contains a mountain known as Mount Sharp, which consists of layers of sedimentary rock that hold clues about the planet's past environments and the possibility of past life. Jezero Crater, on the other hand, will be the landing site for NASA's Perseverance rover in 2021. This site was chosen for its potential to contain a preserved ancient river delta, which could provide evidence of past habitable conditions and the potential for preserved biosignatures.

Real-Life Examples:
Real-life examples of Mars exploration missions, such as the Curiosity rover and the upcoming Perseverance rover mission, showcase the power of 3D mapping and data analysis in unravelling the mysteries of the Red Planet. These missions have provided valuable insights into Mars' geological history, paving the way for future human missions and potential colonization efforts.

Implications for Future Missions:

By studying Mars' geology, topography, and potential exploration sites, scientists can identify areas of interest for future human missions. Understanding the geological composition and history of Mars allows for the identification of resources, such as water and minerals, that could support sustained human presence and exploration on the planet.

Inspiring the Next Generation:
The exploration of Mars ignites curiosity and inspires the next generation of scientists and engineers. Through 3D mapping and the study of Martian geology, young minds are motivated to pursue careers in planetary science and contribute to the future of Mars exploration.

Conclusion:
Mars' 3D representation, with its diverse geological features and potential exploration sites, provides scientists with a rich tapestry of information to unravel the planet's past and its potential for future human missions. By mapping the Martian surface and studying its geology, we gain valuable insights into the planet's history and habitability. The exploration of Mars not only expands our knowledge of the solar system but also sparks our imagination and inspires us to reach for new frontiers.

CHAPTER 12: RED DUST AND EXTREME TEMPERATURES: ENVIRONMENTAL CHALLENGES ON MARS

Introduction:
Mars, often referred to as the Red Planet, presents unique environmental challenges for human exploration. In this chapter, we explore the harsh conditions of Mars, including the pervasive red dust that blankets the planet and the extreme temperatures that fluctuate between cold and scorching heat. Understanding these challenges is crucial for designing habitats, spacesuits, and life support systems that can withstand the Martian environment and ensure the safety and well-being of future human explorers.

The Red Dust of Mars:
One of the defining features of Mars is its red dust, composed mainly of iron oxide particles. The fine dust particles get easily kicked up by the wind, creating vast dust storms that can engulf the entire planet. These storms can pose significant challenges for human missions, as the dust can settle on surfaces, potentially clogging machinery and obstructing solar panels, which rely on

sunlight for power generation. The red dust is also abrasive and can damage equipment, including delicate instruments and electronics.

Real-Life Example: The Martian Dust Storm of 2018
A real-life example of the challenges posed by Martian dust storms is the global dust storm that occurred on Mars in 2018. This storm engulfed the entire planet, significantly reducing sunlight and impacting the operations of NASA's Opportunity Rover. The rover, which relied on solar power, went into a low-power mode to conserve energy during the storm. Although the rover ultimately succumbed to the prolonged lack of sunlight, the dust storm provided valuable data and insights into the behaviour and impact of such events on Mars.

Extreme Temperatures:
Mars is known for its extreme temperature variations. At its equator, temperatures can reach a maximum of around 20 degrees Celsius (68 degrees Fahrenheit) during the day, while plunging to as low as -80 degrees Celsius (-112 degrees Fahrenheit) at night. In the polar regions, temperatures can drop even further, reaching as low as -120 degrees Celsius (-184 degrees Fahrenheit) during the Martian winter. These extreme temperatures pose challenges to human survival, requiring robust thermal insulation and temperature regulation systems to maintain a habitable environment inside habitats and spacesuits.

Real-Life Example: The Curiosity Rover's Thermal Control System
The Curiosity rover, which has been exploring Mars since 2012, utilizes a sophisticated thermal control system to cope with the extreme temperature variations on the planet. The rover is equipped with insulation and heaters to protect its sensitive instruments and electronics from the frigid Martian nights. The thermal control system ensures that the rover remains operational and can withstand the temperature extremes it encounters during its missions.

Mitigating the Environmental Challenges:
To mitigate the challenges posed by the Martian environment, various strategies are being explored. These include the development of dust-resistant materials, improved sealing mechanisms for habitats and spacesuits, and advanced thermal insulation technologies. Additionally, the use of pressurized habitats and controlled environments can provide a stable and habitable space for astronauts, shielding them from harsh external conditions.

Inspiring Resilience and Adaptation:
The environmental challenges of Mars inspire resilience, innovation, and adaptation. Through the exploration and colonization of Mars, humankind is challenged to develop new technologies, materials, and systems that can withstand the hostile Martian environment. The knowledge and experience gained from overcoming these challenges can have far-reaching applications, benefiting not only future Mars missions but also other extreme environments on Earth and beyond.

Conclusion:
Mars presents formidable environmental challenges, from the pervasive red dust to the extreme temperature variations. Understanding and addressing these challenges is essential for successful human exploration and habitation of the Red Planet. Through ongoing research and technological advancements, scientists and engineers are striving to develop robust solutions that can ensure the safety, comfort, and well-being of future astronauts. The exploration of Mars not only pushes the boundaries of human achievement but also serves as a testament to our resilience and adaptability in the face of extreme environments.

PART IV: EXPLORING THE MARTIAN SURFACE

CHAPTER 13: UNVEILING MARS: ROBOTIC ROVERS AND LANDERS

Introduction:
The exploration of Mars has been made possible using robotic rovers and landers, which serve as our eyes and hands on the Red Planet. In this chapter, we delve into the fascinating world of Martian exploration, examining the missions and achievements of these remarkable robotic vehicles. From the pioneering missions of the Viking landers to the more recent successes of the Curiosity rover and the Perseverance rover, we uncover the valuable insights gained from these missions and the technological advancements that have propelled our understanding of Mars.

The Viking Landers: Pioneering the Martian Surface:
The Viking landers, launched by NASA in 1975, were the first robotic missions to successfully land on Mars and conduct experiments on its surface. These missions provided ground-breaking data on the Martian atmosphere, soil composition, and the potential for life. The landers carried a suite of scientific instruments, including cameras, spectrometers, and biology experiments, to study the Red Planet in unprecedented detail.

Real-Life Example: Viking 1 and the Search for Life

Viking 1, one of the twin landers, conducted a series of experiments to search for signs of life on Mars. These experiments included the Labelled Release (LR) experiment, which looked for evidence of metabolic activity by introducing a nutrient solution to Martian soil. Although the results were initially inconclusive, subsequent studies have shed new light on the findings, raising intriguing questions about the possibility of microbial life on Mars.

The Mars Rovers: Expanding Our Understanding:
In the 21st century, a new era of Martian exploration began with the deployment of robotic rovers. These rovers, equipped with advanced scientific instruments and mobility systems, have revolutionized our understanding of Mars. The missions of the Spirit and Opportunity rovers, as well as the more recent Curiosity rover and the Perseverance rover, have provided critical insights into the geology, climate, and potential habitability of the Red Planet.

Real-Life Example: Curiosity's Journey of Discovery
The Curiosity rover, launched in 2011, has been one of the most successful Martian missions to date. It has traversed the Martian surface, exploring the Gale Crater and Mount Sharp. Curiosity's scientific instruments, including a drill for sampling rock and soil, have revealed evidence of past habitable environments and geological processes. The rover's discoveries, such as the detection of organic molecules and the confirmation of an ancient lakebed, have contributed significantly to our understanding of Mars' past and the potential for life.

The Perseverance Rover: Seeking Signs of Life:
The Perseverance rover, launched in 2020, represents the latest milestone in Martian exploration. Equipped with state-of-the-art instruments and technologies, Perseverance aims to search for signs of ancient microbial life, collect samples for future return to Earth, and demonstrate technologies for future human missions. The rover carries the Ingenuity helicopter, which successfully

achieved the first powered flight on another planet, opening new possibilities for aerial exploration.

Real-Life Example: Ingenuity's Historic Flight
The successful flight of the Ingenuity helicopter demonstrated the feasibility of powered flight in the thin Martian atmosphere. This achievement opens new avenues for future exploration, as aerial platforms can cover greater distances and access previously inaccessible areas. The data and images captured by Ingenuity provide valuable insights into the Martian terrain and assist in mission planning for future rovers and human missions.

Conclusion:
Robotic rovers and landers have played a pivotal role in unveiling the mysteries of Mars. These missions have expanded our knowledge of the Red Planet's geology, climate, and potential for life. The successes of the Viking landers, the Mars rovers, and the recent achievements of the Perseverance rover and Ingenuity helicopter demonstrate the remarkable progress in Martian exploration. These missions have paved the way for future human missions and deepened our understanding of our neighbouring planet. The ongoing exploration of Mars holds the promise of unravelling further discoveries and unlocking the secrets of our celestial neighbour.

CHAPTER 14: SUITING UP: EVA (EXTRAVEHICULAR ACTIVITY) ON MARS

Introduction:
Exploring the Martian surface requires astronauts to don specially designed spacesuits for Extravehicular Activity (EVA). In this chapter, we delve into the complexities of suiting up for Mars missions, examining the challenges of operating in the harsh Martian environment and the innovative technologies incorporated into Martian spacesuits. We explore the importance of EVA in conducting scientific research, collecting samples, and maintaining equipment, as well as the training and preparations required for astronauts to safely venture outside their habitats.

The Martian Environment:
Before discussing Martian spacesuits, it is crucial to understand the unique conditions on the Red Planet. Mars has a thin atmosphere, extreme temperature variations, and a surface exposed to harmful radiation. These factors pose significant challenges for human exploration and necessitate the development of highly advanced and protective spacesuits.

Real-Life Example: The Mars Dune Buggy Incident
During a simulated Mars mission in the Utah desert, a team of analogue astronauts encountered difficulties when their rover

became stuck in a dune. The astronauts, wearing spacesuits designed for Mars-like conditions, had to employ their EVA skills to free the rover and continue their mission. This example underscores the need for well-designed and reliable spacesuits to ensure the safety and success of future Martian missions.

The Evolution of Martian Spacesuits:
Spacesuit technology has come a long way since the early days of space exploration. Martian spacesuits must provide a life-sustaining environment for astronauts, protecting them from the thin atmosphere, extreme temperatures, and harmful radiation. They must also allow for freedom of movement and dexterity to perform complex tasks. Over the years, advancements in materials, mobility systems, and life support systems have been made to meet the unique demands of Mars exploration.

Real-Life Example: The Z-2 Prototype Spacesuit
NASA's Z-2 prototype spacesuit showcases the cutting-edge technology being developed for future Martian missions. It features a hard upper torso for protection, advanced mobility joints for increased flexibility and improved radiation shielding. The Z-2 prototype demonstrates the continuous refinement of spacesuit design and the commitment to ensuring the safety and comfort of astronauts in the challenging Martian environment.

EVA Procedures and Preparations:
EVA operations on Mars require meticulous planning and preparation. Astronauts must undergo extensive training in EVA protocols, including practising various scenarios and emergency procedures. They must also be well-versed in the operation of their spacesuits and the tools and equipment they will use during EVAs. Communication and coordination among the EVA team members are critical for a successful and safe mission.

Real-Life Example: The Apollo 17 Moonwalk
While not on Mars, the Apollo 17 moonwalk serves as a significant example of the importance of EVA procedures and

preparations. Astronauts Eugene Cernan and Harrison Schmitt conducted three moonwalks, collecting samples, and performing scientific experiments. Their successful EVAs demonstrated the value of well-planned and executed extravehicular activities in maximizing scientific returns and achieving mission objectives.

Future Innovations in Martian Spacesuits:
As we look to future Mars missions, ongoing research and development are focused on further improving spacesuit technology. This includes advancements in materials, such as self-healing fabrics and enhanced radiation protection, as well as innovations in mobility and life support systems. These advancements will contribute to the continued safety, efficiency, and comfort of astronauts during EVAs on Mars.

Conclusion:
Extravehicular Activity is a critical component of human exploration on Mars. Spacesuits designed for the Martian environment are essential for protecting astronauts and enabling them to conduct scientific research, explore the surface, and carry out maintenance tasks. The evolution of Martian spacesuits, from the early days of space exploration to the advanced prototypes of today, demonstrates our dedication to overcoming the challenges of operating in the harsh Martian environment. With ongoing advancements and preparations, we are inching closer to the day when astronauts will step foot on Mars, supported by state-of-the-art spacesuits that enable them to explore the Red Planet safely and effectively.

CHAPTER 15: SEARCHING FOR LIFE: ASTROBIOLOGY AND THE QUEST FOR MARTIAN BIOSIGNATURES

Introduction:
The search for extra-terrestrial life has captivated scientists and space enthusiasts for decades. In this chapter, we delve into the field of astrobiology and its role in the exploration of Mars. We explore the fascinating possibility of finding evidence of past or present life on the Red Planet and the scientific methods employed to detect and study Martian biosignatures. From the tantalizing discovery of water on Mars to the ongoing exploration missions, the quest for signs of life on Mars continues to ignite our imagination and reshape our understanding of the universe.

The Martian Environment and Habitability:
Before discussing the search for Martian biosignatures, it is crucial to understand the environmental conditions on Mars and its potential for supporting life. Mars, with its thin atmosphere, extreme temperatures, and radiation exposure, presents significant challenges for life as we know it. However,

recent discoveries of liquid water, subsurface ice, and the presence of organic molecules have sparked optimism about the potential habitability of Mars.

Real-Life Example: The Curiosity Rover's Discoveries
NASA's Curiosity rover, launched in 2011, has been instrumental in our understanding of Mars' habitability. Its exploration of Gale Crater has revealed ancient lake beds, evidence of flowing water, and organic compounds. These findings provide valuable insights into the potential for past habitable environments on Mars and fuel the search for biosignatures.

Astrobiology and the Search for Biosignatures:
Astrobiology is an interdisciplinary field that combines biology, chemistry, geology, and astronomy to study the origin, evolution, and distribution of life in the universe. When it comes to Mars, astrobiologists focus on identifying biosignatures—indirect evidence or traces of life—through various scientific methods. These methods include studying rock formations, analysing soil and atmospheric samples, and utilizing advanced imaging and spectroscopic techniques.

Real-Life Example: The Viking Missions
The Viking missions, conducted by NASA in the 1970s, were the first to search for signs of life on Mars. The landers performed experiments to detect metabolic activity and conducted chemical analyses. Although the results were inconclusive at the time, the Viking missions laid the foundation for future Astro biological investigations on Mars.

Instruments and Techniques for Detecting Biosignatures:
To search for biosignatures on Mars, scientists utilize a suite of sophisticated instruments and techniques. These include mass spectrometers, gas analysers, DNA sequencers, and microscopes. Additionally, the deployment of robotic explorers, such as rovers and landers, allows for in-situ analysis of Martian samples.

Real-Life Example: The Mars Sample Return Mission

NASA and the European Space Agency (ESA) are planning a Mars Sample Return mission, aiming to collect Martian samples and return them to Earth for detailed analysis. This ambitious mission holds great potential for discovering definitive evidence of past or present life on Mars and revolutionizing our understanding of the Martian biosphere.

Challenges and Future Prospects:
The search for biosignatures on Mars faces numerous challenges, including the potential contamination of Martian samples, the limitations of remote sensing techniques, and the complexities of interpreting data. However, advancements in technology and future exploration missions, such as NASA's Perseverance rover and the ESA's ExoMars rover, offer new opportunities for breakthrough discoveries.

Conclusion:
The quest for Martian biosignatures is a captivating journey of scientific exploration and discovery. While we have not yet found conclusive evidence of past or present life on Mars, the ongoing missions and advancements in astrobiology continue to inspire and fuel our curiosity about the possibility of extra-terrestrial life. The search for biosignatures on Mars not only expands our understanding of the potential for life beyond Earth but also highlights the interconnectedness of the universe and the profound questions that drive our exploration of the cosmos.

CHAPTER 16: RESOURCE UTILIZATION ON MARS: IN-SITU RESOURCE EXTRACTION AND UTILIZATION (ISRU)

Introduction:
As humans embark on the journey to Mars, the ability to sustain life and conduct long-duration missions becomes paramount. One of the key challenges is the availability of essential resources in a hostile and inhospitable environment. In this chapter, we explore the concept of In-situ Resource Utilization (ISRU) and its crucial role in enabling human exploration and colonization of Mars. We delve into the potential resources on Mars, the technologies for extracting and utilizing them, and the implications for sustainable and self-sufficient missions.

Understanding Martian Resources:
Mars is rich in resources that could be utilized to support human activities. These resources include water, which can be found

as ice in the polar caps and subsurface, and carbon dioxide in the Martian atmosphere. Additionally, Martian soil, known as regolith, contains elements such as iron, aluminium, silicon, and oxygen, which can be extracted and used for various purposes.

Real-Life Example: Mars Ice Home
NASA's Mars Ice Home concept is an innovative design that utilizes Martian water ice as a structural component for habitats. By extracting and processing the ice, it could provide not only water for drinking and agriculture but also shielding from radiation and insulation against extreme temperatures.

In-situ Resource Extraction:
To utilize Martian resources, advanced technologies are required for their extraction and processing. These technologies may include drilling and excavation techniques, water extraction methods, and chemical processes to extract usable materials from the regolith. The goal is to establish sustainable resource utilization systems that minimize reliance on Earth for resupply.

Real-Life Example: Mars Oxygen In-Situ Resource Utilization Experiment (MOXIE)
NASA's Perseverance rover, equipped with the Mars Oxygen In-Situ Resource Utilization Experiment (MOXIE), aims to demonstrate the production of oxygen from the Martian atmosphere. This technology is crucial for future human missions as it can potentially produce breathable air and serve as a propellant for rockets, reducing the need for Earth-based supplies.

Utilizing Martian Resources:
Once resources are extracted, they can be utilized in various ways to support human activities on Mars. Water can be used for drinking, agriculture, and the production of rocket propellant through electrolysis. Carbon dioxide can be processed to generate oxygen and methane, which can serve as fuel for return missions or energy sources for power systems.

Real-Life Example: 3D Printing with Regolith

Scientists and engineers are exploring the possibility of using Martian regolith as a raw material for 3D printing. This technology would enable the construction of habitats, infrastructure, and tools directly on Mars, reducing the need for transporting bulky materials from Earth.

Implications for Sustainable and Self-Sufficient Missions:
The successful implementation of ISRU technologies on Mars has significant implications for long-term human presence and colonization. By reducing the dependence on Earth for essential resources, missions become more sustainable and self-sufficient. This not only lowers costs but also increases the resilience and adaptability of future Martian colonies.

Conclusion:
In-situ Resource Extraction and Utilization (ISRU) is a fundamental concept for enabling human exploration and colonization of Mars. By leveraging the resources available on the Red Planet, we can reduce the reliance on Earth and establish self-sustaining missions. The ongoing research and development in ISRU technologies pave the way for a future where humans can thrive on Mars, utilizing its resources to support scientific exploration, economic activities, and the eventual expansion of our presence in the solar system.

PART V: CHALLENGES AND RISKS OF MARTIAN EXPLORATION

CHAPTER 17: RADIATION AND SPACE WEATHER: PROTECTING ASTRONAUTS FROM COSMIC HAZARDS

Introduction:
As humans venture beyond Earth's protective atmosphere and journey into space, they face the challenges posed by cosmic radiation and space weather. In this chapter, we explore the nature of radiation in space, the effects it has on astronauts, and the measures taken to protect them during long-duration missions. We delve into the types of radiation encountered in space, the risks they pose, and the strategies employed to mitigate those risks and ensure the safety and well-being of astronauts.

Understanding Space Radiation:
Space is filled with various forms of radiation, including galactic cosmic rays (GCRs), solar particle events (SPEs), and trapped radiation in planetary magnetic fields. GCRs consist of high-energy particles originating from outside our solar system, while SPEs are intense bursts of radiation emitted by the Sun during solar flares. Trapped radiation, such as the Van Allen Belts around

Earth, poses additional hazards.

Real-Life Example: Apollo 11 and Solar Particle Events
During the Apollo 11 mission, astronauts Neil Armstrong, Buzz Aldrin, and Michael Collins encountered a solar particle event. To mitigate the risks, the crew received instructions to turn their spacecraft, the Columbia, so that the lunar module, the Eagle, acted as a shield against the radiation. This precautionary measure helped protect the astronauts during their historic mission.

Radiation Risks and Health Effects:
Exposure to space radiation poses several health risks to astronauts. The primary concern is the potential for developing cancer due to the cumulative effect of radiation exposure over time. Additionally, radiation can cause acute effects such as radiation sickness and damage to organs and tissues, including the central nervous system, cardiovascular system, and reproductive system.

Real-Life Example: International Space Station (ISS) and Radiation Monitoring
The International Space Station (ISS) is equipped with various radiation monitoring instruments to measure the radiation dose received by astronauts in real time. These measurements help scientists understand the radiation environment and assess the potential risks to astronauts' health.

Mitigating Radiation Risks:
To protect astronauts from radiation, several mitigation strategies are employed. Shielding materials, such as aluminium and polyethene, are used in spacecraft construction to attenuate radiation. Mission planning considers solar activity forecasts to minimize exposure during solar particle events. Astronauts also adhere to strict radiation safety protocols, including time limits for spacewalks and designated shelter areas during radiation events.

Real-Life Example: Orion Spacecraft and Radiation Shielding
NASA's Orion spacecraft, designed for future human exploration missions, incorporates advanced radiation shielding technologies. The spacecraft's crew module is equipped with radiation-shielding materials and water-filled bags that act as additional shielding against cosmic radiation.

Advancements in Radiation Research:
Ongoing research is focused on developing improved radiation protection strategies and understanding the long-term effects of space radiation on human health. This includes studying the use of pharmaceutical agents, genetic modifications, and regenerative medicine approaches to mitigate radiation damage and enhance the body's natural defence mechanisms.

Real-Life Example: NASA's Twins Study
NASA's Twins Study, involving astronaut Scott Kelly and his identical twin brother Mark Kelly, provided valuable insights into the effects of long-duration spaceflight on the human body, including radiation exposure. The study revealed changes in gene expression, DNA damage, and immune system function, furthering our understanding of the risks associated with space radiation.

Conclusion:
Radiation and space weather present significant challenges for human space exploration. By understanding the nature of space radiation, and its risks, and employing effective mitigation strategies, we can protect astronauts during their missions beyond Earth. Ongoing research and advancements in radiation protection techniques will continue to enhance the safety and well-being of future astronauts, paving the way for further exploration and the eventual colonization of space.

CHAPTER 18: PSYCHOLOGICAL CONSIDERATIONS: COPING WITH ISOLATION AND THE CHALLENGES OF MARS

Introduction:
The journey to Mars and the prospect of living on the red planet for extended periods pose unique psychological challenges for astronauts. Isolation, confinement, distance from Earth, and the demanding nature of space travel can have profound effects on the mental well-being of individuals. In this chapter, we explore the psychological considerations and strategies employed to help astronauts cope with isolation and overcome the challenges they may face during missions to Mars.

Understanding the Psychological Challenges:
Space missions to Mars involve long-duration space travel, with astronauts spending months or even years away from Earth. This isolation, coupled with the confined living spaces of spacecraft and the lack of direct communication with loved ones, can lead to feelings of loneliness, homesickness, and psychological stress. Additionally, the inherent risks and uncertainties of space

exploration can contribute to anxiety and other psychological issues.

Real-Life Example: The Mars500 Experiment
The Mars500 experiment, conducted by the European Space Agency (ESA) and the Russian Institute for Biomedical Problems, aimed to simulate a mission to Mars. Six volunteers were confined in a simulated spacecraft for 520 days to study the psychological and physiological effects of long-duration isolation. The findings provided valuable insights into the psychological challenges faced by astronauts on future Mars missions.

Psychological Support Systems:
To address the psychological challenges of Mars missions, psychological support systems are put in place. Astronauts undergo extensive psychological evaluations and training before their missions. They receive support from mission control and have access to communication tools for regular contact with their families and a support network on Earth. Psychologists and mental health professionals also provide ongoing support and counselling throughout the mission.

Real-Life Example: The International Space Station (ISS) and Crew Support
The International Space Station (ISS) serves as a model for understanding and addressing the psychological well-being of astronauts. Crew members on the ISS have access to a variety of support mechanisms, including regular video conferences with family members and scheduled leisure time to engage in activities that help alleviate feelings of isolation.

Cognitive and Behavioural Strategies:
Astronauts are trained in cognitive and behavioural strategies to help them cope with the challenges of isolation and confinement. These strategies include stress management techniques, problem-solving skills, maintaining a daily routine, and engaging in physical exercise and recreational activities. Maintaining a

positive outlook, fostering camaraderie among crew members, and cultivating a sense of purpose and mission also play vital roles in maintaining psychological well-being.

Real-Life Example: Chris Hadfield and Psychological Resilience
Canadian astronaut Chris Hadfield's experience on the ISS serves as an inspiring example of psychological resilience in space. Through his social media presence and engaging videos, Hadfield demonstrated how maintaining a positive mindset, staying connected with Earth, and embracing the challenges of space exploration can contribute to psychological well-being during long-duration missions.

Conclusion:
The psychological well-being of astronauts on Mars missions is a critical aspect of ensuring their overall health and success. By understanding the unique psychological challenges posed by isolation and confinement, implementing effective support systems, and providing training in cognitive and behavioural strategies, astronauts can develop the resilience and coping mechanisms necessary to thrive in the demanding environment of Mars. The lessons learned from real-life examples and research will continue to inform the development of comprehensive psychological support systems for future Mars missions, enabling astronauts to overcome challenges and embark on this remarkable frontier of human exploration.

CHAPTER 19: PLANETARY PROTECTION: PREVENTING CONTAMINATION OF MARS AND EARTH

Introduction:
As humanity prepares to embark on missions to Mars, it is crucial to consider the potential impact of our presence on the Martian environment and the potential contamination of Earth with extra-terrestrial organisms. Planetary protection aims to minimize the risk of biological contamination and preserve the scientific integrity of both Mars and our home planet. In this chapter, we explore the importance of planetary protection, the protocols in place, and the real-life examples that highlight its significance.

Preserving Mars: The Case for Planetary Protection:
Mars holds great scientific value as a potentially habitable planet and a potential host for past or present life. It is essential to preserve the pristine nature of Mars to conduct accurate studies and search for signs of life without interference from terrestrial contamination. Planetary protection measures are in

place to prevent spacecraft and landers from introducing Earth's microorganisms, which could compromise scientific discoveries and hinder our understanding of the Martian ecosystem.

Real-Life Example: Viking Missions
The Viking missions, launched by NASA in the 1970s, were the first to land on Mars and conducted experiments to search for signs of life. Although the results were inconclusive, they highlighted the importance of planetary protection protocols. The Viking missions implemented strict cleanliness requirements to reduce the chances of contamination and ensure the accuracy of the experiments.

Planetary Protection Policies and Guidelines:
International bodies, such as the Committee on Space Research (COSPAR), have established guidelines and policies for planetary protection. These policies categorize missions based on their level of contamination risk and outline the necessary precautions and sterilization procedures. The goal is to prevent the unintentional introduction of Earth organisms that may interfere with Mars' scientific investigation or pose potential risks to indigenous Martian life.

Real-Life Example: Mars Rover Missions
The Mars rover missions, including the Curiosity rover and the upcoming Perseverance rover, exemplify the stringent planetary protection protocols implemented by space agencies. Before launch, these rovers undergo extensive sterilization processes to eliminate or reduce the number of microorganisms carried from Earth. This ensures the integrity of scientific investigations on Mars and helps safeguard against potential contamination.

Challenges and Controversies:
Implementing effective planetary protection measures presents challenges and controversies. Some argue that overly stringent protocols may hinder exploration and scientific progress, while others emphasize the need for utmost caution to protect both

planets. Balancing scientific exploration and planetary protection requires ongoing discussions and adaptations to ensure the best possible outcomes for both Mars and Earth.

Real-Life Example: Mars Sample Return Mission
The planned Mars Sample Return mission, a collaboration between NASA and the European Space Agency (ESA), poses unique challenges for planetary protection. Bringing Martian samples back to Earth requires meticulous planning to prevent any potential contamination of Earth's biosphere. Stringent containment measures and isolation protocols will be implemented to ensure the safe handling and study of Martian materials.

Conclusion:
Planetary protection is essential to safeguard Mars' pristine environment and prevent potential contamination of Earth. Through adherence to established guidelines and protocols, space agencies and scientific communities demonstrate their commitment to preserving the integrity of both planets. The real-life examples of past missions and the ongoing discussions surrounding future endeavours highlight the continuous efforts to strike a balance between scientific exploration and the protection of our planetary ecosystems. By prioritizing planetary protection, we can enhance our understanding of Mars while maintaining the delicate balance of our biosphere.

CHAPTER 20: EMERGENCY SITUATIONS ON MARS: MITIGATING RISKS AND ENSURING CREW SAFETY

Introduction:
As humans venture to Mars, they face the inherent risks and challenges of exploring an unforgiving environment millions of kilometres away from Earth. Emergencies can arise unexpectedly, posing threats to crew safety and mission success. In this chapter, we delve into the importance of preparedness, the potential emergency scenarios on Mars, and the measures taken to mitigate risks and ensure the safety of the crew.

Understanding the Risks:
Mars presents several hazards that could lead to emergencies, including extreme temperatures, dust storms, radiation exposure, and the possibility of technical failures in life support systems or habitats. Each of these risks must be carefully considered in mission planning to develop effective emergency response strategies.

Real-Life Example: Apollo 13 Mission
While not on Mars, the Apollo 13 mission serves as a compelling example of how emergencies can arise during space exploration. An oxygen tank explosion crippled the spacecraft, leaving the crew in a life-threatening situation. Through resourcefulness, teamwork, and the ingenuity of mission control, the crew safely returned to Earth. This mission highlights the importance of preparedness and adaptability in handling emergencies.

Emergency Response Protocols:
Mission planners and space agencies develop comprehensive emergency response protocols to address potential crises on Mars. These protocols include procedures for evacuations, medical emergencies, equipment failures, communication breakdowns, and contingency plans for prolonged stays or isolation. Crew members receive rigorous training to prepare them for various emergency scenarios.

Real-Life Example: Mars Desert Research Station (MDRS)
To simulate Mars missions and test emergency response procedures, analogue habitats like the Mars Desert Research Station in Utah have been established. These simulations provide valuable insights into how astronauts would handle emergencies on Mars, allowing for the refinement of protocols and the identification of potential weaknesses in current systems.

Communication and Resupply:
Maintaining communication between Mars and Earth is vital during emergencies. Redundant communication systems and protocols are established to ensure timely and effective communication with mission control and support teams on Earth. Additionally, resupply missions play a crucial role in providing necessary resources, medical supplies, and equipment to sustain the crew and address emergencies.

Real-Life Example: International Space Station (ISS)
The International Space Station serves as a testing ground for

emergency response procedures and crew safety protocols. With continuous human presence in space, the ISS has faced numerous challenges, including ammonia leaks, fire outbreaks, and medical emergencies. Each incident has provided valuable lessons that inform emergency response plans for future Mars missions.

Psychological and Emotional Well-being:
In emergencies, the psychological and emotional well-being of the crew becomes paramount. Crew members must be equipped with effective coping strategies, stress management techniques, and support systems to navigate stressful and potentially life-threatening circumstances. Psychological evaluations, counselling services, and regular communication with loved ones are crucial to maintaining mental health during emergencies.

Real-Life Example: Antarctic Research Stations
Antarctic research stations, where teams endure extreme isolation and challenging conditions, offer insights into the psychological challenges of long-duration missions. The strategies developed to support the mental well-being of researchers in Antarctica can inform the psychological support systems for astronauts on Mars missions.

Conclusion:
Emergencies pose significant challenges during Mars missions, requiring meticulous planning, comprehensive protocols, and ongoing training. Real-life examples from past space missions and analogue simulations on Earth provide valuable lessons for addressing emergencies on Mars. By prioritizing preparedness, communication, crew safety, and psychological well-being, we can enhance the chances of successfully navigating and mitigating risks during emergencies, ensuring the safety and success of human missions to Mars.

PART VI: THE FUTURE OF MARTIAN EXPLORATION

CHAPTER 21: BEYOND MARS: MARS AS A STEPPINGSTONE TO THE OUTER SOLAR SYSTEM

Introduction:
While Mars is a captivating destination for human exploration, it also serves as a steppingstone for venturing further into the outer reaches of our solar system. In this chapter, we explore the potential of Mars as a launch point for future missions to outer planets, such as Jupiter, Saturn, Uranus, and Neptune. We discuss the advantages of using Mars as a staging area, the challenges associated with extended space travel, and the possibilities of unlocking the mysteries of our cosmic neighbourhood.

Utilizing Mars as a Launch Point:
Mars possesses several advantages that make it an ideal location for launching missions deeper into the solar system. Its relatively low gravity and atmosphere provide opportunities for efficient fuel utilization and aerobraking, reducing the energy requirements for interplanetary travel. By leveraging Mars as a launch point, spacecraft can conserve fuel and achieve higher velocities, enabling faster and more cost-effective journeys to the outer planets.

Real-Life Example: Voyager Missions

The Voyager 1 and Voyager 2 missions serve as remarkable demonstrations of utilizing planetary flybys to reach the outer solar system. Launched in 1977, these spacecrafts used gravity assists from Jupiter and Saturn to propel themselves toward Uranus, Neptune, and beyond. The success of the Voyager missions highlights the significance of leveraging celestial bodies for efficient interplanetary travel.

Challenges of Extended Space Travel:

Journeying to the outer planets presents unique challenges due to the vast distances, prolonged mission durations, and the need for long-term life support systems. Astronauts must contend with increased radiation exposure, extended isolation, and the physical and psychological effects of prolonged weightlessness. Developing technologies and strategies to mitigate these challenges is crucial for safe and successful missions.

Real-Life Example: New Horizons Mission

The New Horizons mission, launched in 2006, exemplifies the challenges and triumphs of long-duration space travel. After a nine-year journey, the spacecraft conducted a historic flyby of Pluto, providing valuable insights into this enigmatic dwarf planet and its moons. The success of the New Horizons mission demonstrates our ability to overcome the challenges of extended space travel and unlock the mysteries of distant celestial bodies.

Exploring the Outer Planets:

The outer planets, such as Jupiter, Saturn, Uranus, and Neptune, hold immense scientific value and offer unparalleled opportunities for exploration. These gas giants and ice giants possess unique atmospheres, diverse moons, and enigmatic magnetic fields that can provide valuable insights into the formation and evolution of our solar system. Missions to these planets can help us unravel their mysteries and deepen our understanding of planetary science.

Real-Life Example: Cassini-Huygens Mission
The Cassini-Huygens mission, a joint venture between NASA, the European Space Agency (ESA), and the Italian Space Agency (ASI), provided an unprecedented view of Saturn and its moons. The mission not only studied the planet's magnificent rings but also revealed intriguing details about its moons, including the presence of liquid oceans beneath their surfaces. The success of the Cassini-Huygens mission showcases the scientific discoveries that can be made through missions to the outer planets.

Prospects and Inspiring Discoveries:
Looking ahead, Mars can serve as a launch point for ambitious missions to the outer planets, offering scientists and explorers the opportunity to unravel the mysteries of our cosmic neighbourhood. These missions could provide insights into the formation of the solar system, the potential for life beyond Earth, and the dynamics of planetary atmospheres. They inspire humanity to push the boundaries of exploration and deepen our understanding of the universe.

Conclusion:
Mars, with its advantageous position and resources, acts as a crucial steppingstone for human exploration and scientific investigations of the outer solar system. By leveraging Mars as a launch point, we can optimize our interplanetary missions, conserve resources, and propel ourselves toward the outer planets and beyond. As we continue to explore the mysteries of our cosmic neighbourhood, Mars stands as a gateway to inspiring discoveries and the realization of our aspirations for interplanetary exploration.

CHAPTER 22:
MARS AND HUMAN COLONIZATION: BUILDING A SUSTAINABLE FUTURE ON THE RED PLANET

Introduction:
In this chapter, we delve into the possibilities of human colonization on Mars and the prospects of establishing a sustainable future on the Red Planet. We explore the challenges and considerations involved in creating a self-sufficient Martian colony, from habitat design and resource utilization to food production and long-term sustainability. By envisioning a future where humans call Mars home, we unlock the potential for interplanetary civilization and the expansion of humanity beyond Earth.

The Vision of a Martian Colony:
Human colonization of Mars represents a bold and ambitious endeavour, driven by the desire to explore, expand our horizons, and ensure the long-term survival of our species. Establishing a self-sustaining colony on Mars requires careful planning, robust infrastructure, and innovative technologies to overcome the

harsh Martian environment and create a thriving community.

Real-Life Example: Mars One Project
The Mars One project, initiated in 2011, captured the imagination of the public by proposing a plan to establish a permanent human settlement on Mars by 2032. Although the project faced financial and technical challenges and was ultimately discontinued, it highlighted the growing interest in Martian colonization and the belief in the feasibility of creating a sustainable human presence on the Red Planet.

Habitat Design and Construction:
Designing habitats suitable for long-term living on Mars is a critical aspect of colonization efforts. Structures must protect against extreme temperatures, radiation, and the thin Martian atmosphere. Utilizing local resources, such as Martian regolith and ice, can aid in the construction of habitats and reduce reliance on Earth-supplied materials.

Real-Life Example: NASA's 3D-Printed Habitat Challenge
NASA's 3D-Printed Habitat Challenge encouraged teams to design and build habitats using autonomous construction technologies and resources found on Mars. This challenge aimed to push the boundaries of habitat design and construction methods, showcasing innovative solutions for future Martian colonies.

Resource Utilization and Self-Sufficiency:
To ensure long-term sustainability, Martian colonies must maximize resource utilization and achieve self-sufficiency. In-situ resource utilization (ISRU) technologies, such as extracting water from Martian ice and utilizing the carbon dioxide-rich atmosphere for oxygen production, can help reduce the reliance on Earth for essential resources.

Real-Life Example: Mars Oxygen In-Situ Resource Utilization Experiment (MOXIE)
The Mars Oxygen In-Situ Resource Utilization Experiment (MOXIE) aboard NASA's Perseverance rover demonstrates the

feasibility of producing oxygen from the Martian atmosphere. By converting carbon dioxide into oxygen, MOXIE showcases a crucial step toward achieving self-sufficiency in future Martian colonies.

Food Production and Agricultural Systems:
Establishing sustainable food production systems on Mars is vital for the long-term survival of colonists. Hydroponics, aeroponics, and other innovative agricultural methods can be employed to cultivate crops in controlled environments, providing essential nutrition and psychological well-being to the Martian settlers.

Real-Life Example: The International Space Station (ISS) Veggie Project
The Veggie project aboard the International Space Station demonstrates the potential of growing fresh food in space using plant growth chambers. This project provides valuable insights into the challenges and benefits of space agriculture, informing future efforts to develop sustainable food production systems on Mars.

Long-Term Sustainability and Adaptation:
The success of a Martian colony relies on long-term sustainability and adaptability. Technologies for waste recycling, energy generation, and environmental monitoring must be implemented to create a closed-loop ecosystem that minimizes waste and maximizes resource efficiency.

Real-Life Example: The BIOS-3 Experiment
The BIOS-3 experiment conducted in Russia from 1965 to 1991 simulated a closed ecosystem that provided oxygen, food, and water for a crew of three individuals. This experiment showcased the potential for creating self-sustaining habitats and informed our understanding of long-term sustainability in isolated environments.

Conclusion:
The colonization of Mars represents a monumental undertaking

that pushes the boundaries of human exploration and ingenuity. By envisioning a sustainable future on the Red Planet, we open possibilities for interplanetary civilization and ensure the survival of our species in the face of potential challenges on Earth. Through habitat design, resource utilization, and long-term sustainability, we lay the groundwork for a thriving Martian colony that serves as a testament to human resilience, adaptability, and the indomitable spirit of exploration.

CHAPTER 23: MARS AND INTERPLANETARY COOPERATION: INTERNATIONAL COLLABORATION IN MARTIAN EXPLORATION

Introduction:
In this chapter, we explore the importance of international collaboration in the exploration and colonization of Mars. The exploration of Mars is a global endeavour that requires the expertise, resources, and cooperation of multiple nations. By working together, countries can pool their knowledge and resources, share costs, and accelerate progress in Martian exploration. In this chapter, we delve into the history of international collaboration in Mars missions, discuss the benefits of such cooperation, and highlight notable examples of countries joining forces to explore the Red Planet.

The Need for International Collaboration:
Mars exploration is a complex and expensive endeavour

that demands the collective efforts of multiple nations. By collaborating, countries can leverage their unique strengths, share scientific data, and contribute to a greater understanding of Mars and its potential for human habitation. International cooperation also helps mitigate the risks and costs associated with Mars missions, ensuring a more efficient and sustainable exploration program.

Real-Life Example: International Mars Exploration Working Group (IMEWG)
The International Mars Exploration Working Group, comprising space agencies from around the world, fosters international collaboration in Mars exploration. Through regular meetings and cooperation agreements, IMEWG facilitates information sharing, joint mission planning, and the exchange of scientific data, enhancing global efforts in Martian exploration.

Sharing Resources and Expertise:
International collaboration allows participating countries to share their resources, including launch capabilities, scientific instruments, and research facilities. By pooling resources, countries can reduce costs and increase the scope of their missions, enabling more comprehensive exploration and data collection.

Real-Life Example: NASA and ESA ExoMars Collaboration
NASA and the European Space Agency (ESA) have collaborated on the ExoMars mission, which aims to search for signs of past or present life on Mars. This collaboration combines the expertise and resources of both agencies, allowing for the deployment of advanced scientific instruments and increasing the chances of success in discovering Martian life.

Joint Missions and Cooperative Projects:
Countries often join forces to undertake joint missions to Mars, combining their scientific and technological capabilities to achieve shared objectives. These missions involve the

sharing of spacecraft, instruments, and scientific data, fostering greater scientific collaboration and enabling more comprehensive exploration.

Real-Life Example: NASA's Mars Science Laboratory (MSL) Curiosity Rover
The Curiosity rover, a joint mission between NASA and international partners, exemplifies the power of international cooperation in Mars exploration. The rover, equipped with state-of-the-art scientific instruments, has been instrumental in studying the Martian environment, providing valuable insights into the planet's geology, climate, and potential for habitability.

Promoting Peaceful Cooperation and Diplomacy:
International collaboration in Mars exploration promotes peaceful cooperation and diplomacy among nations. By working together toward a common goal, countries build stronger relationships, foster mutual understanding, and transcend geopolitical boundaries. This collaborative spirit paves the way for future joint missions and scientific discoveries.

Real-Life Example: The United Arab Emirates Hope Probe
The United Arab Emirates Hope Probe mission to Mars is a testament to the power of international collaboration. By partnering with international entities and institutions, the UAE has developed its space capabilities and positioned itself as a regional leader in space exploration. This mission not only advances scientific knowledge but also promotes peace and cooperation in the Middle East region.

Conclusion:
International collaboration in Martian exploration is essential for the progress and success of future missions to Mars. By pooling resources, expertise, and knowledge, countries can overcome the challenges of space exploration, enhance scientific discoveries, and work toward a shared vision of understanding and colonizing the Red Planet. Through collaborative efforts, we can unlock

the mysteries of Mars and pave the way for the future of interplanetary exploration.

CHAPTER 24:
DREAMS OF MARS: INSPIRING THE NEXT GENERATION OF EXPLORERS

Introduction:
In this final chapter, we explore the power of Mars exploration to inspire and ignite the dreams of future generations. The quest to explore Mars captivates the human imagination, encouraging innovation, curiosity, and a sense of adventure. By sharing the wonders of Martian exploration and showcasing the possibilities of interplanetary travel, we can inspire the next generation of explorers and scientists who will shape the future of Mars missions and beyond.

Inspiring through Education and Outreach:
Education plays a crucial role in inspiring young minds to pursue careers in science, technology, engineering, and mathematics (STEM) fields. Mars exploration offers a captivating backdrop to engage students and encourage their interest in space exploration. Through educational programs, hands-on activities, and interactive exhibits, we can spark the curiosity of young minds and cultivate a passion for exploration.

Real-Life Example: Mars Student Imaging Project (MSIP)

The Mars Student Imaging Project, led by NASA, enables students to request images of Mars taken by the Mars Reconnaissance Orbiter's HiRISE camera. This initiative provides students with the opportunity to actively participate in scientific research, fostering a deeper understanding of Mars and encouraging a passion for space exploration.

Celebrating Exploration Heroes:
Real-life exploration heroes, such as astronauts and scientists, serve as role models and sources of inspiration for aspiring explorers. By highlighting the achievements and stories of these individuals, we can ignite the dreams and aspirations of young minds, showing them that anything is possible with determination and perseverance.

Real-Life Example: Dr Mae Jemison
Dr Mae Jemison, the first African American woman to travel to space, serves as an inspiration to countless individuals around the world. Her pioneering spirit and commitment to exploration inspire young people to reach for the stars and pursue their dreams, regardless of their background or gender.

Collaborating with Entertainment Media:
Entertainment media, including books, movies, and documentaries, have the power to captivate audiences and ignite their imagination. By incorporating Mars exploration narratives into popular culture, we can inspire a wide range of people and create a sense of wonder and excitement about the possibilities of space exploration.

Real-Life Example: The Martian (Book and Film)
"The Martian," a novel by Andy Weir and its subsequent film adaptation, captured the public's imagination with its thrilling story of survival and ingenuity on Mars. The narrative showcased the challenges and triumphs of human exploration, inspiring audiences around the world to consider the possibilities of interplanetary travel.

Promoting Diversity and Inclusion:
Inspiring the next generation of explorers involves promoting diversity and inclusion in the space sector. By creating an inclusive and equitable environment, we can ensure that all individuals, regardless of their background, can contribute to Mars exploration and shape the future of space exploration.

Real-Life Example: Artemis Program and Lunar Gateway
NASA's Artemis program, which aims to return humans to the Moon, emphasizes the importance of diversity and inclusion in space exploration. The program includes plans to send the first woman and the first person of colour to the lunar surface, inspiring a new generation of explorers and showcasing the importance of representation in space missions.

Conclusion:
Mars, with its mysteries and potential for human exploration, has the power to ignite the dreams and aspirations of future generations. By leveraging education, celebrating exploration heroes, collaborating with entertainment media, and promoting diversity and inclusion, we can inspire the next generation of explorers who will lead humanity's journey to Mars and beyond. Let us continue to nurture their dreams, provide them with the tools and resources they need, and empower them to reach for the stars.

www.ingramcontent.com/pod-product-compliance
Lightning Source LLC
Chambersburg PA
CBHW062358290526
45794CB00005B/2279